Eddy the Electron Goes

Solar

A fun and educational story about photovoltaics

Kim Keahiolalo

Published by Auberson & Graydon Productions, LLC
Santa Fe, New Mexico

Author: Kim Keahiolalo
Illustrator: Blaise Auberson
Art Director: Steve Graydon

Visit www.solareddy.com for upcoming publications.

Certificate of Registration for Eddy VAu 1-001-153
Effective date of registration September 20, 2009
For Phyllis VAu 1-001-150 September 20, 2009

In 1921 Albert Einstein was awarded the Nobel Prize in Physics for his simple explanation of how light can be absorbed by an element. When an element such as silicon absorbs light from the sun, it will emit electrons.

Einstein's work led to the development of modern solar cells. Solar cells convert light into electricity, also known as the photovoltaic effect. The photovoltaic effect was first demonstrated over 100 years ago.

This is a story about how a photon named Phyllis collided with an electron named Eddy and the adventures that follow their exciting encounter.

Eddy had just moved into a brand new solar panel
in the desert. He liked his new place a lot and
the weather outside was really great.
Nice and sunny!

Little did he know what was about to happen
next would change his life forever.

From 93 million miles away, Phyllis the Photon came flying down from her home in the sun. She was headed straight for Eddy's house in the solar panel.

Phyllis was moving so fast and had so much energy that when she got to the solar panel she knocked Eddy right out of his cozy orbital. He was now free to float around the photovoltaic cell which was one great big silicon crystal.

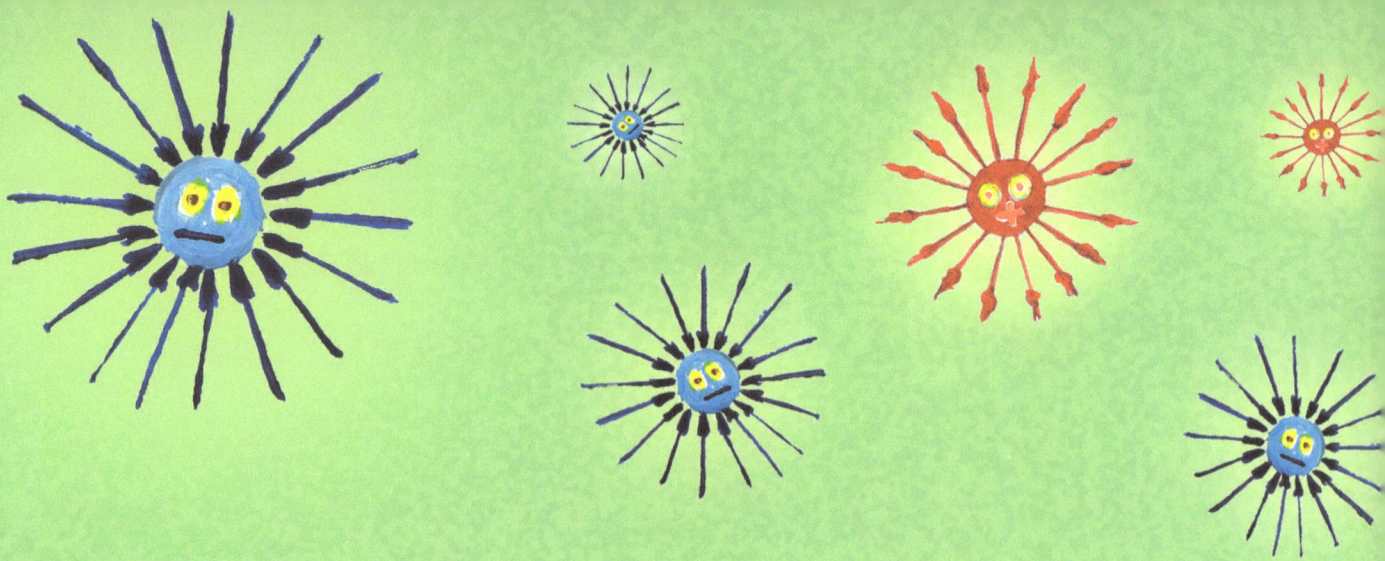

Eddy was so excited, he barely knew what to do with all his energy. But as he was floating free in the silicon, he felt himself suddenly drawn to one side! The designers of the photovoltaic cell had built in a p-n junction down the middle, which created a permanent electric field that would show electrons the way. Without any more bouncing around, Eddy knew exactly where to go: straight to the edge of the cell, where a copper wire was already beckoning!

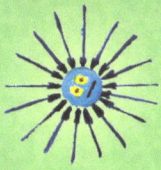

On his way, he flew past some holes. They were lonely atoms, missing one of their electrons. The holes were moving too! Some electrons on their way past would fall into these holes and settle back into cozy orbitals, just like their old home atom. Eddy felt a little bit sorry for the lonely atoms, but he knew they wouldn't stay empty for long. Eddy arrived at the copper terminal with energy to spare.

Eddy realized he was no longer in his solar panel. Although this was all new, Eddy felt surprisingly positive about his future. He was as free as a bird. He jumped on a jet-ski and surfed down the current on the copper wires.

Up ahead was a strange looking building—it was called the inverter. The other electrons around him told Eddy that the inverter is where the direct current gets changed to alternating current.

When Eddy got to the front door
of the inverter he jumped off his jet-ski.
He had so much energy that he decided he would
let off some steam by doing a little shadow boxing.

He thought it was also a good idea to lift some weights before going into the inverter. That way he would be strong for the next part of his journey. Eddy likes to be prepared!

The inverter was a very strange and confusing place. Eddy and the electrons around him watched as the wire in front of them disappeared. There suddenly seemed no place to go. Then a path opened up, but it wasn't clear where it went. Eddy started to move ... but now a path appeared on the other side. Then it was gone again! There! Gone! Other side! And it was all happening so quickly.

Eddy realized the inverter was teaching him to do the alternating current dance. With a little practice, it started to make sense. Eddy swung back and forth, back and forth, to the groovy beat of 60 cycles every second. Dancing with the other electrons, Eddy now made a wave that could actually travel out the other side of the inverter! He was now alternating current.

Eddy was really having a blast. He was on the electric grid and there were billions and billions of other electrons there too—all surfing on nice smooth sine waves.

Eddy was being pushed along by the high voltage from all the other electrons on the grid. It was a surfer's delight.

Eddy traveled like this for miles and miles. His thoughts turned back to Phyllis and how she had come down from the sun and changed his life. Eddy knew that without Phyllis he would still be hanging out in his silicon atom, spinning around in his cozy orbital.

Eddy was beginning to wonder where it was he was headed on this big electric grid. All this activity was starting to wear him down and he wanted a break from surfing the high tension wires.

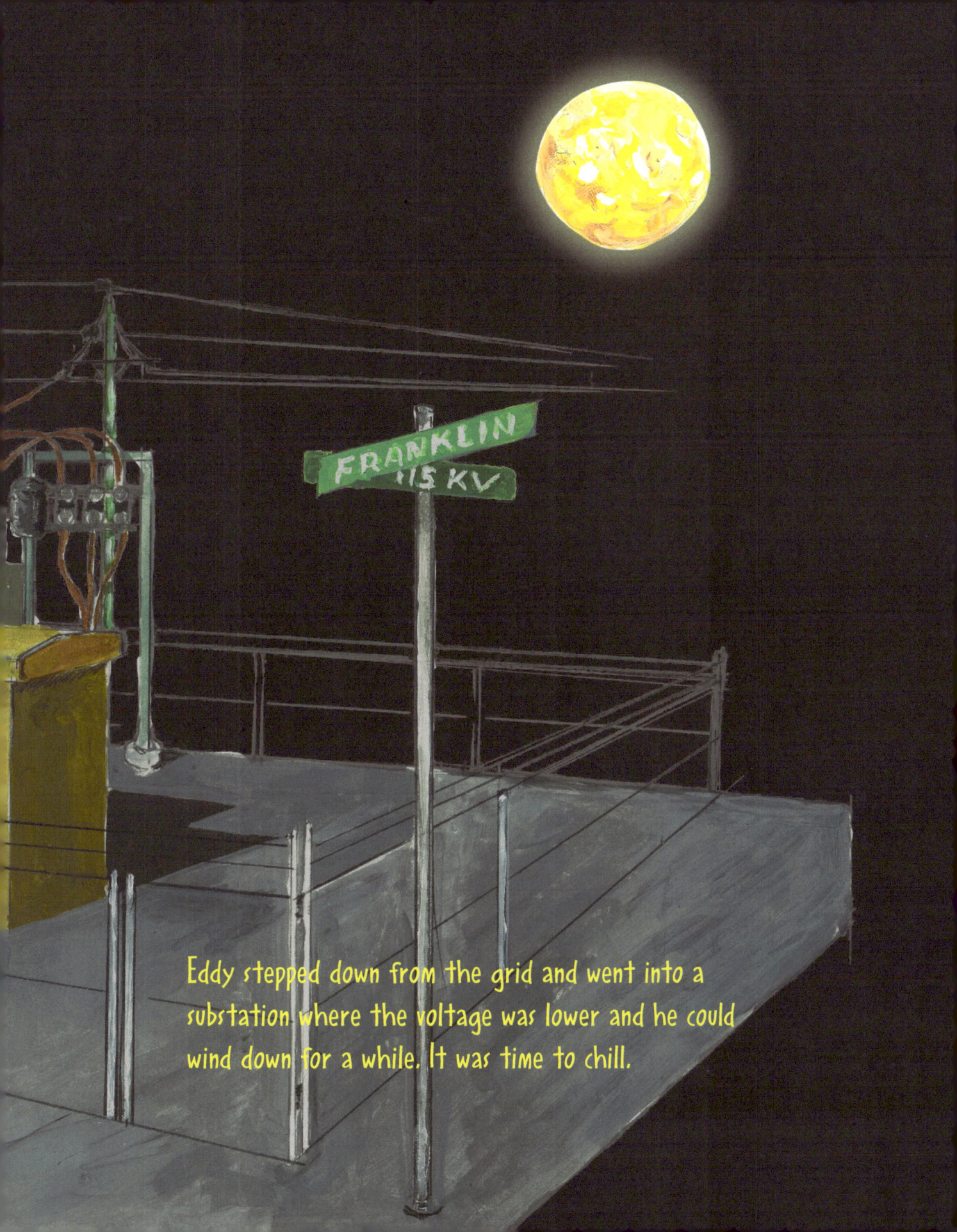

Eddy stepped down from the grid and went into a substation where the voltage was lower and he could wind down for a while. It was time to chill.

Then SPLAT!!! Without any warning Eddy was being pulled through a utility meter. What could possibly be next?

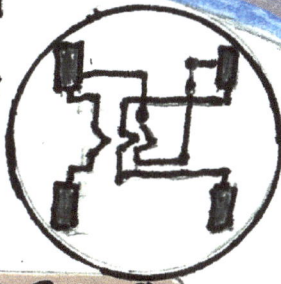

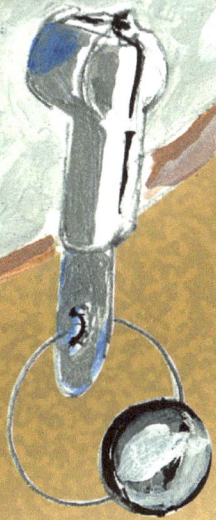

Eddy's destiny was unfolding. He had been knocked loose from his orbital by Phyllis so that he could share his energy with the world.

Eddy now has a fabulous job in the city. He works for an electric train company. He helps thousands of people move around everyday in a clean and efficient style.

Clean electrons are beginning to be used everywhere. Many of Eddy's electron friends have very important jobs. They power lights, motors, pumps and fans, to mention a few.

They are providing environmentally friendly electricity for our homes, farms, schools, and all manner of enterprises and transportation. Eddy hopes that clean electrons will soon be available to power the world of the future.

Eddy appreciates Phyllis for coming down from
the sun and sharing her light so that he could
be transformed into useful electricity.

Watts=

W=VI

amperes × volts

Eddy is also very thankful to all the scientists and engineers who worked to develop solar energy. He hopes we will continue to make solar energy more efficient and cost-effective. Everybody should have access to solar power!

The -
End

Glossary of Terms

Ampere (A)	A unit of electric current. One ampere refers to a specific number of electrons passing a fixed point each second.
Atom	The smallest unit of matter indivisible by chemical means.
Bi-directional Meter	A meter that measures electrical energy over time whether the energy is flowing to the electrical grid or from the grid.
Current	The flow of electricity through a wire.
Diode	An electronic component that conducts electric current in only one direction.
Einstein, Albert	(1879–1955) Einstein was a theoretical physicist, philosopher and author who is widely regarded as one of the most influential and best known scientists and intellectuals of all time. He is often regarded as the father of modern physics.
Electric Field	A property that describes the space that surrounds electrically charged particles. An electric field exerts a force on other electrically charged objects.
Electron	One of the parts of the atom having a negative charge. Indivisible particle with a charge of –1.
Energy	The ability to make something happen or to do work.
Grid	An electricity delivery system from point of generation to point of consumption.
Inverter	An electrical converter that changes direct current (DC) into alternating current (AC).
Meter	An electricity-measuring device. An electric meter measures the flow of energy from a distribution line to a building or other end-use over time.
Nobel Prize	Any of the six international prizes awarded annually by the Nobel Foundation for outstanding achievements in the fields of physics, chemistry, physiology or medicine, literature, and economics and for the promotion of world peace.
Orbital	A place where electrons live, sometimes in pairs and sometimes not.
Photon	An elementary particle that is the basic unit of light. A tiny packet of energy.
Photovoltaic	The word "photovoltaic" combines two terms — "photo" means light and "voltaic" means voltage. Photovoltaic cells directly convert sunlight into electricity.
p-n Junction	The basis of an electronic device called a diode, which allows electric current to flow in only one direction.
Power	The rate at which work is performed or energy is converted.
Silicon	A semiconductor, readily either donating or sharing its four outer electrons allowing many different forms of chemical bonding. A chemical element, which has the symbol Si and atomic number 14. In Earth's crust, silicon is the second most abundant element after oxygen.
Sine wave	A continuous, uniform wave with a constant frequency and amplitude.
Solar	Produced by or coming from the sun.
Solar panel	Also known as a module; it is a collection of solar cells. A solar cell is any device that directly converts the energy in light (photons) into electrical energy through the process of photovoltaics.
Substation	Part of the electrical grid. The point or location where high voltage transmission of electricity is transformed to lower and safer voltage for electrical distribution to a meter.
Volt (V)	A unit of measurement of force, or pressure, in an electrical circuit.
Watt (W)	Named after James Watt (1736–1819). The standard unit of measurement of electrical power. One watt is one ampere of current flowing at one volt.

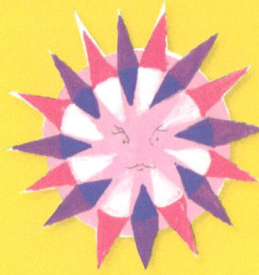

Eddy the Electron is a collaborative effort between long time friends,
Kim Keahiolalo, Blaise Auberson and Steve Graydon.

The Author

Kim Keahiolalo earned a B.A. in Environmental Studies and Planning from Sonoma State
University in 2005. Kim focused her education on Energy Management and Design —
a curriculum of engineering, architecture and economics. Kim would like to thank her
teachers, Mr. Galen George and Dr. Alexandra von Meier for making learning about
chemistry and energy a fun and valuable experience.

The Illustrator

Swiss-born artist Blaise Pascal Auberson is a graduate of Silvermine College of Art in
New Canaan, Connecticut. He has had many art exhibits in Europe and the US. Blaise
works in several mediums including painting, sculpture and mixed media. He spent most
of his life in Northern California working as an artist and a residential green builder.

The Art Director

Graphics designer Steve Graydon has studied art in California and Aix-en-Provence,
France. He hopes America and the world will soon find their way to a clean energy future.

A sample of upcoming titles from Auberson & Graydon Productions, LLC

Eddy and Sweeney Visit the Wetlands

Eddy Researches the Smart Grid

Eddy Explores Biomass Energy Sources

Eddy Catches the Wind

Eddy Digs Geothermal Energy

Eddy Dives Into Hydropower

This book is dedicated to the children of the world.

Thank you to Dr. von Meier for her joyful contributions to the book regarding the p–n junction and the inverter.

www.ingramcontent.com/pod-product-compliance
Lightning Source LLC
LaVergne TN
LVHW072109070426
835509LV00002B/85